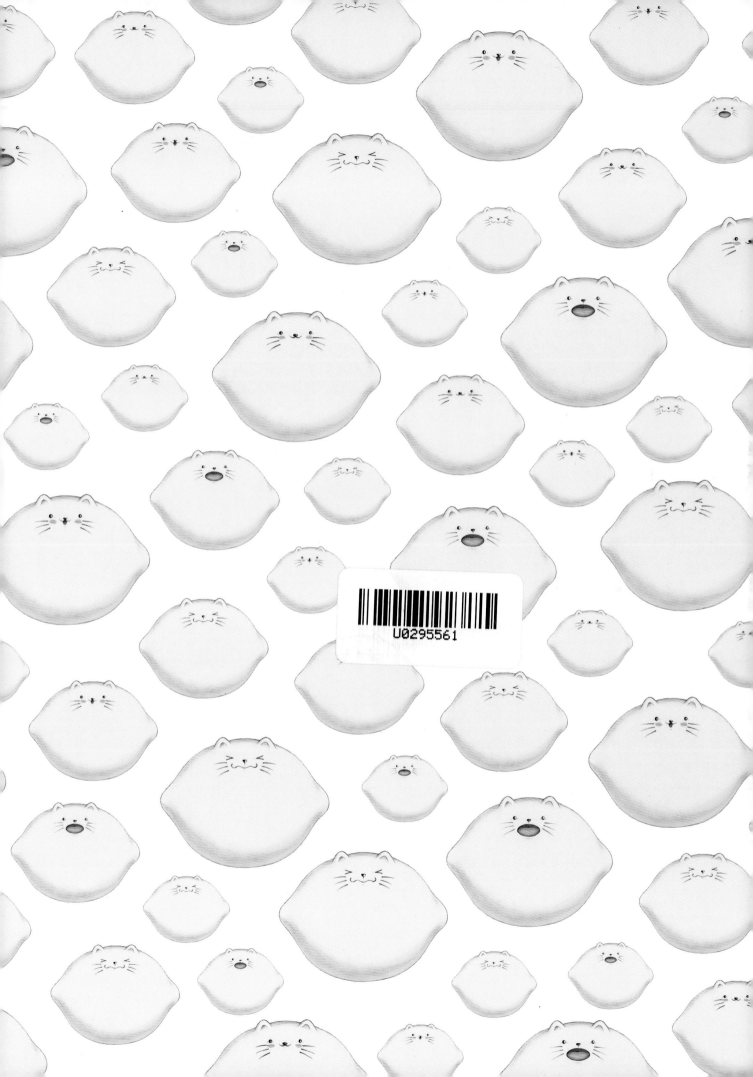

图书在版编目 (CIP) 数据

猫柠猫奇遇记之美洲丛林 / 陈楠工作室著绘 . —上海：
上海三联书店, 2019.1
　　ISBN 978-7-5426-6514-0

　　Ⅰ.①猫… Ⅱ.①陈… Ⅲ.①热带雨林—美洲—儿童
读物 Ⅳ.① S717.1-49

中国版本图书馆 CIP 数据核字 (2018) 第 238464 号

猫柠猫奇遇记之美洲丛林

著 · 绘：陈楠工作室
责任编辑：陈马东方月
监　制：姚　军
责任校对：戚智轩
装帧设计：陈楠工作室

出版发行：上海三联书店
　　　　　(200030) 中国上海市漕溪北路 331 号 A 座 6 楼
邮购电话：021-22895540
印　　刷：上海雅昌艺术印刷有限公司

版　次：2019 年 1 月第 1 版
印　次：2019 年 1 月第 1 次印刷
开　本：889×1194　1/16
字　数：2 千字
印　张：3
书　号：ISBN 978-7-5426-6514-0/J.285
定　价：39.80 元

敬启读者，如发现本书有印装质量问题，请与印刷厂联系 021-66798999

猫柠猫奇遇记 之 美洲丛林

Morningmore's American Jungle Adventure

著/绘：陈楠工作室

上海三联书店

绘本馆里住着一只像柠檬一样的黄色大胖猫,叫猫柠猫。

每天晚上,当绘本馆里的小朋友都离开后,猫柠猫便会在静悄悄的绘本馆里,找他最心爱的绘本——《文字秘籍》。

There's a chubby lemon cat, called Morningmore, living in a picture book shop.

Every night, after everyone leaves, Morningmore will go straight to his favorite picture book "Mystical Book".

《文字秘籍》被猫柠猫藏在绘本馆最深处的书架上。这是猫柠猫的探险家爷爷传给他的探险笔记。他多么想像爷爷一样做个探险家呀。

"Mystical Book" is always hidden by Morningmore in the deepest corner of the shop. It is an adventure journal, passed to him by his adventurer grandfather. Morningmore has been dreaming of becoming an adventurer just like his grandpa.

今天，他翻到了新的一页，上面画着奇怪的树丛和一些古怪的符号。他正仔细端详着，突然发现其中一个由四瓣叶子组成的符号好像幽幽地发着金光。于是，他好奇地伸出手，在那符号上摸了一摸……

Tonight, Morningmore turns to a new page with strange jungle and mystical symbols on it. While reading it, he finds a four-leaf clover glowing in a golden light. He reaches out to it……

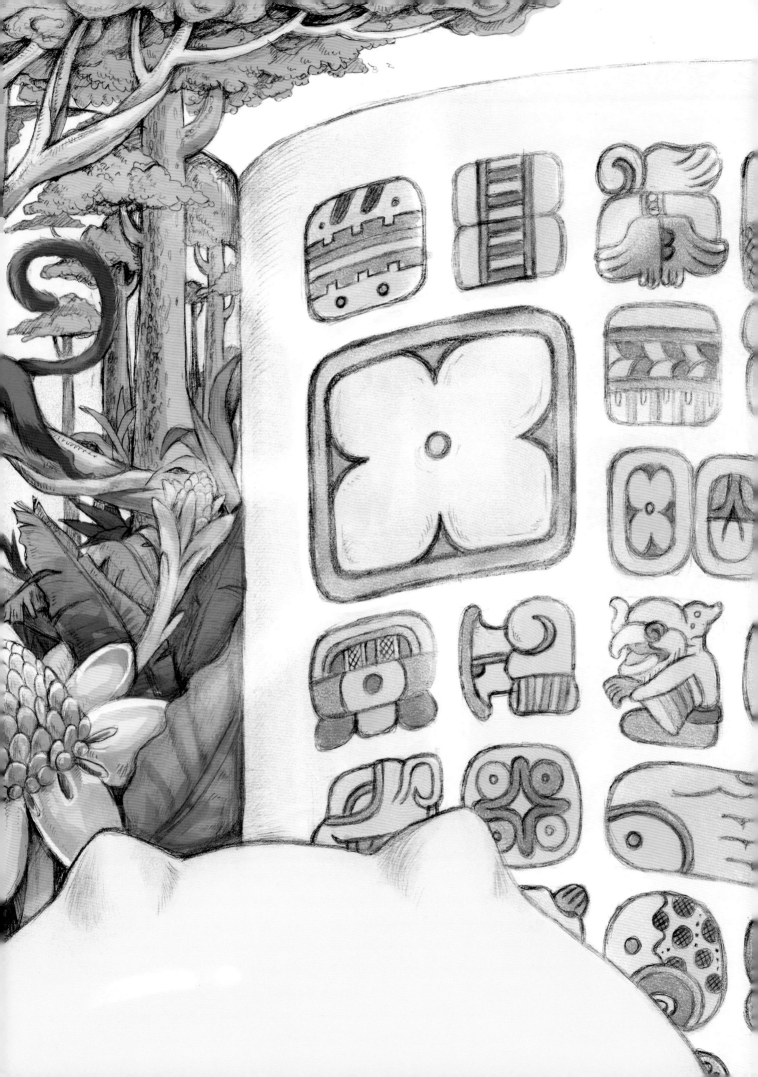

"嗖"的一声,猫柠猫被吸入了书中的世界。

"啊……!救命啊!"猫柠猫拼命地叫。

"不要怕,你很安全。"一片巨大的叶子温柔地接住了他。

"谢谢你救了我,你是谁?"

"我是大根乃拉草(1),世界上草本植物中叶子最大的植物!欢迎来到美洲丛林(2)!"

小贴士:所有带标注的小知识,都可以在本书末尾的《猫柠猫秘密图鉴》里找到哦!

"Whoosh," Morningmore rolls into the world of "Mystical Book".

"Help! Help!" Morningmore screams.

"Calm down, you are safe," a huge leaf catches him gently.

"Thanks for saving me. Who are you?"

"I am Giant Rhubarb (1), the largest herbaceous plant in the world! Welcome to the American Jungle (2) !"

"美洲在哪？美洲丛林又是什么？"猫柠猫问道，向周围张望。这时，一棵树从猫柠猫眼前跑过。

"咦？树怎么会跳啊？"

"我是行走的棕榈树（3），我的根就是我的腿，每当季节变化时，我就会追随着阳光前进！你要看看热带雨林吗？跟我来！"

"哇，太好啦！可以开始探险了！"猫柠猫一跃而下，跟着行走的棕榈树跑进了丛林深处。

"Where is it?" Morningmore is deep in wonder, looking around for an answer. At the moment, a tree runs past him.

"What? How can a tree run?"

"I am Socratea (3), one of the spices of palms. My roots are my legs and as the seasons change, I run around following the sunlight. Would you like to take a trip? Follow me!"

"Wow, that's fantastic! Let's begin our adventure!" Morningmore takes a big leap and follows Socratea into the jungle.

树林深处,粗壮的树木盘根错节,巨大而芬芳的花朵处处盛开,还有各种从没见过的鸟和小动物,欢迎着这位远道而来的客人。猫柠猫又惊又喜,"一定能交到很多好朋友!"

Deep in the jungle, Morningmore sees this kind of view for the first time. Big and sweet flowers are everywhere. Huge trees with damp smells are covering the sky. And chirping birds are flying around. They are all welcoming Morningmore who comes from far away.

"I definitely can make a lot of friends!" Morningmore is surprised and excited.

"呱~~~呱~~~你从哪里来？怎么从没见过你？"一个声音冒了出来。

"啊，我叫猫柠猫，是小小的探险家！你是谁？你在哪？"猫柠猫一边说一边找。

"我是箭毒蛙(4)，是个毒药大师！别看我在青蛙家族里个头最小，可是毒性第一，连毒蜘蛛都是我的盘中餐呢！不过看你挺可爱，咱们俩做朋友吧，我是不会欺负朋友的。"

"啊，荣幸荣幸！"猫柠猫连忙答应，真不敢相信自己在这里的第一个朋友，竟是个身怀绝技的小不点。

A frog is croaking "where are you from? I have never seen you before." "My name is Morningmore. I am an adventurer! Who are you and where are you? I can't find you."

"I am Poison Dart Frog (4), master of poison, the smallest, but the most poisonous frog. Even the spiders are easy prey to me. However, you look so cute and let's be friends. I always take good care of my friends."

"Nice to meet you" Morningmore replies. He can't believe his first friend is this little fella with enormous energy.

"嗡~~~嗡~~~"一个细细的声音传来,"你是谁?怎么从没见过你?"

"我叫猫柠猫,是小小探险家,你是谁?你在哪?"猫柠猫四处张望。

"我是蜂鸟(5),世界上最小的鸟!我的喙像一根针,我的舌头像一丝线,我能吃到花朵最深处的花蜜!"

"呀,原来你在花丛里呢!"猫柠猫细细观察,才发现和他说话的小鸟竟然和蜜蜂一样小。

蜂鸟扇动着翅膀说:"都说友谊就像蜜一样甜,咱们俩做好朋友吧!"

"Buzzing……Buzzing……," a weak sound comes through, "who are you? I've never seen you before."

"My name is Morningmore. I am an adventurer! Who are you and where are you?" Morningmore looks around.

"I'm Hummingbird (5), the smallest bird in the world. My beak is like a needle and my tongue is like a thread, which allow me to eat the nectar in furthest end of the flowers."

"Oh, you are here, among the flowers!" Morningmore looks closely and find out that hummingbirds hum just like bees.

Humming birds flap their wings rapidly. "Friendship is sweet as honey. Let's be good friends."

"扑~~~扑~~"一阵翅膀扇动的声音从四面八方围拢过来。

猫柠猫抬头一看:"啊,是蝴蝶!"他惊叫出来。

"你好,我是88多涡蛱蝶(6),我的翅膀上天生长着人类的数字88!""你好,我是光明女神蝶(7),人们用希腊美神的名字将我命名!""你好,我是红带袖蝶(8),我的种族已经在地球上存在了数百万年!"

"你们好美啊!"猫柠猫简直不敢相信自己的眼睛。

"和我们一起跳舞吧!"蝴蝶们形成了一股彩色的旋风,把猫柠猫带到了空中,飞啊飞,越飞越高,最后将他放在了一棵大树上。"

小朋友们,你们在图上找到这三种蝴蝶了吗?可以翻到书的后面,对照答案看看你们找的对不对~

"Hoof……hoof……" the unceasing and innumerable rustle of wings is coming from every direction. Morningmore looks up and sees lots of butterflies.

"Hello, my name is Anna's Eight-eight (6), which comes from the number 88 on my wings!"

"I'm Morpho Helena (7), named after Greek Goddess of Beauty."

"Hi, I'm Postman Butterfly (8). My species has been living on Earth for millions of years!"

"You are all so pretty!" Morningmore can't believe his eyes.

"Let's dance." Butterflies fly around, forming a colorful tornado, bringing Morningmore to the sky, higher and higher. Morningmore finally steps onto the top of a huge tree.

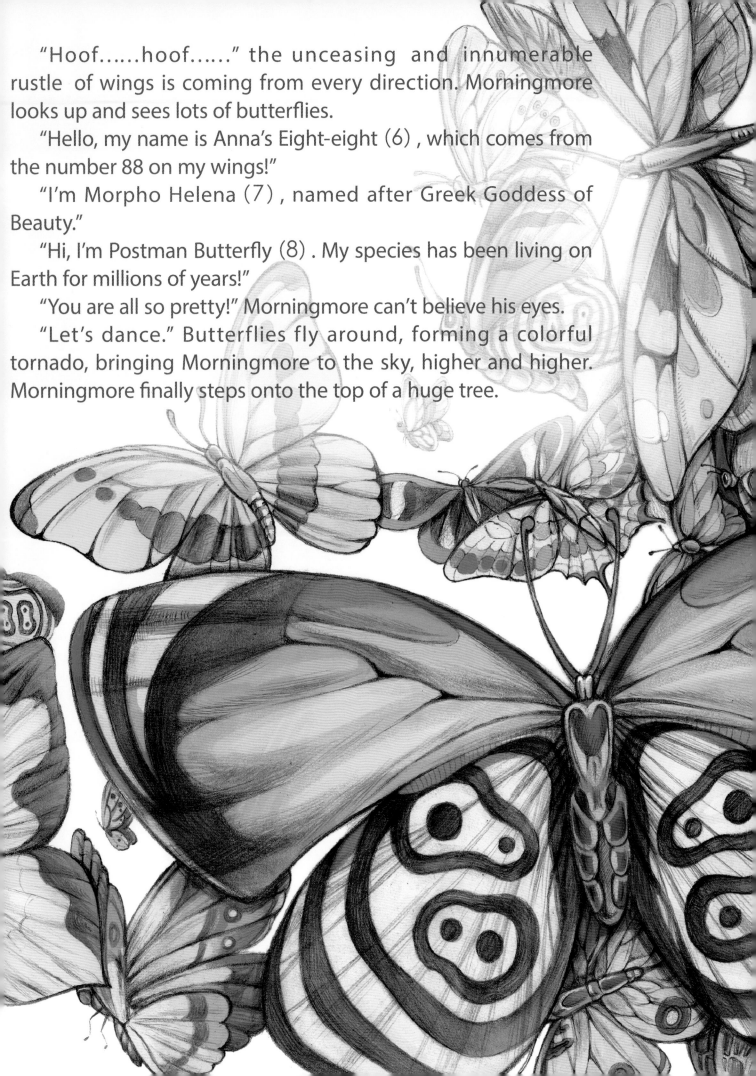

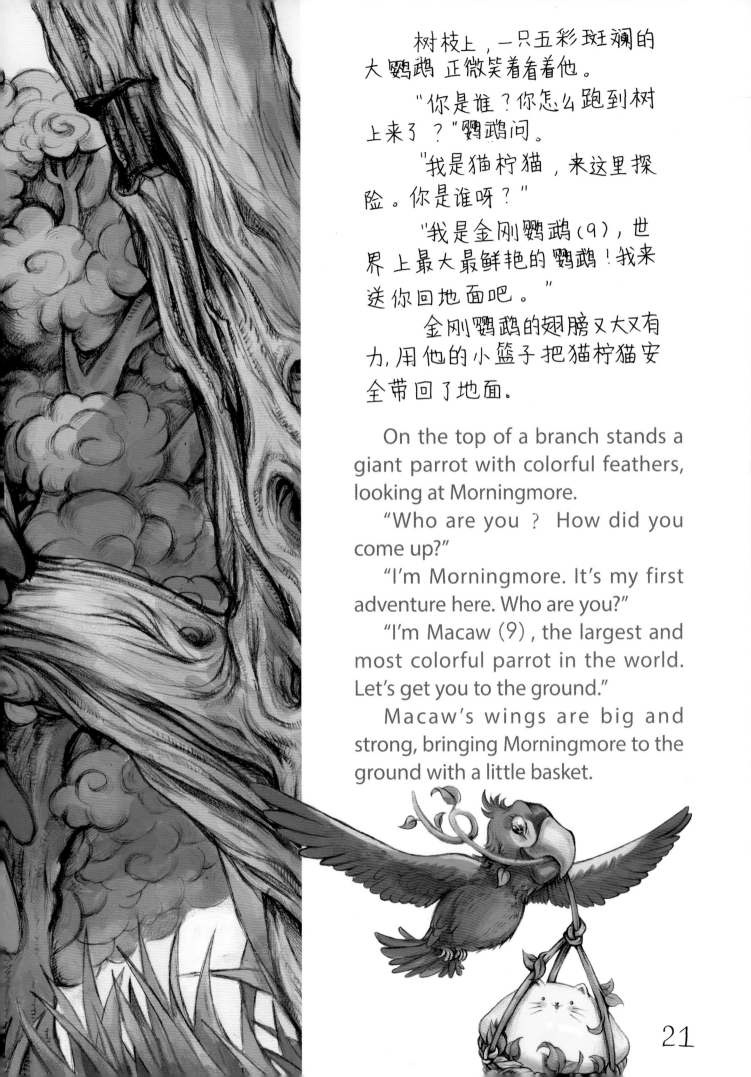

树枝上，一只五彩斑斓的大鹦鹉正微笑着看着他。

"你是谁？你怎么跑到树上来了？"鹦鹉问。

"我是猫柠猫，来这里探险。你是谁呀？"

"我是金刚鹦鹉（9），世界上最大最鲜艳的鹦鹉！我来送你回地面吧。"

金刚鹦鹉的翅膀又大又有力，用他的小篮子把猫柠猫安全带回了地面。

On the top of a branch stands a giant parrot with colorful feathers, looking at Morningmore.

"Who are you？ How did you come up?"

"I'm Morningmore. It's my first adventure here. Who are you?"

"I'm Macaw (9), the largest and most colorful parrot in the world. Let's get you to the ground."

Macaw's wings are big and strong, bringing Morningmore to the ground with a little basket.

"谢谢你啦!"告别了金刚鹦鹉,猫柠猫继续向河边走去。

"哗啦啦,哗啦啦……"草丛里突然传出一阵令人不安的嘈杂声。

猫柠猫一回头,竟有一张巨大的嘴从草丛后面伸了出来。

"嘿嘿,好肥一只猫!正好做我凯门鳄(10)的美餐!"

猫柠猫吓坏了,他从没见过这么大的嘴,这么多的尖牙。可他又肥又圆的身子怎么来得及跑?猫柠猫急中生智,对鳄鱼叫道:"别吃我!我,我,我不是猫,我是柠檬!"

"Thanks a lot!" After farewell to Macaw, Morningmore walks towards the river.

"Walala……Walala……" a disturbing noise is coming out of the bush. When Morningmore turns around, a gigantic mouth is reaching out of the bush.

"Aha, a big fat cat, such a wonderful meal for Caiman (10) ."

Morningmore is freaking out. He has never seen such a big mouth and so many sharp teeth, but he is too fat to run away. Morningmore comes up with a brilliant idea. He says to the crocodile "don't eat me, I……I……I'm not a cat. I'm a lemon."

"看我的,柠檬汁喷发!"

"啊,我的眼睛!"凯门鳄被酸酸的柠檬汁刺得睁不开眼,他哇哇乱叫:"胆敢和我这亚马逊河最凶猛的霸主作对!你等着瞧!"

猫柠猫赶紧把身体缩成一个球,向丛林深处滚去。

"Look here. Lemon juice power!"

"Oh no, my eyes!" Caiman's eyes are burnt by that highly acidic juice. He screams out loud "how dare you to fight against the fiercest animal in this jungle. You'll see!"

At the moment, Morningmore rolls away like a ball.

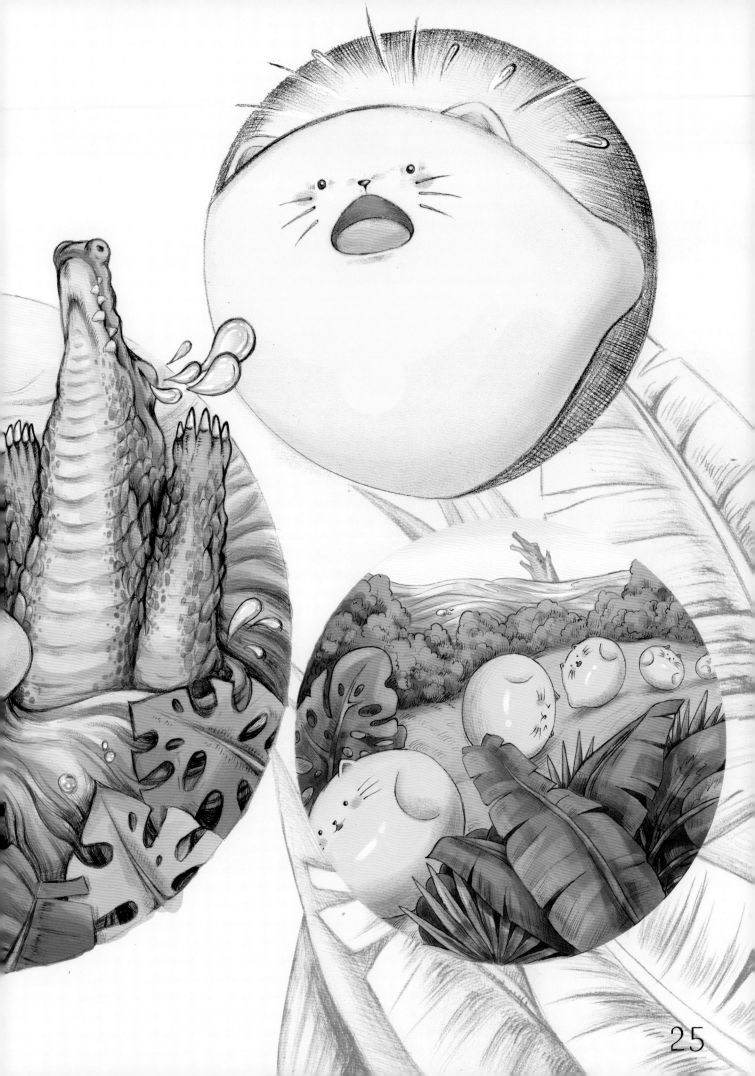

猫柠猫滚啊滚,凯门鳄的嚎叫声还在从身后阵阵传来,猫柠猫叫道:"救命啊!救命啊!"

"你是谁啊?怎么啦?"树林后面突然出现了一个清亮的声音。

猫柠猫一抬头,一个皮肤黑乎乎,头上插满羽毛的小男孩正看着他。

"我叫猫柠猫,有只鳄鱼在追我!"

"猫柠猫,别害怕!我叫玛雅,这片森林的每一个角落我都清清楚楚,我来帮助你!"

Morningmore keeps rolling and he can still hear the voice of Caiman coming after him. He keeps screaming "Help! Help!"

"Who are you and what happened?" A voice comes out of the jungle.

Morningmore looks up and sees a dark-skinned boy with leathers decorating his hair. Morningmore explains "my name is Morningmore and there's a Caiman coming after me."

"Don't be afraid. My name is Maya. I know this place like the back of my hand. I'll help you."

With Maya's loud whistle, a big pin[k] bird comes out of nowhere and flie[s] towards them. It's a flamingo. They rid[e] the flamingo (12), flying higher an[d] higher. The flamingo's pink feather is a[s] beautiful as afterglow. They fly over th[e] Victoria Amazonica (13), which are eve[n] bigger than the bed Morningmore sleep[s] in. They take a break in the jungle an[d] wonder how big those fruits are on th[e] trees.

"These are cocoa trees (14) an[d] those fruits with sharp ends are coco[a] beans. The drink made with coco[a] beans is a little bit bitter but aromati[c,] extremely delicious."

"So, these are cocoa beans. We mak[e] chocolate with them."

玛雅向空中吹了一声口哨，竟有一只粉红色的大鸟飞了过来。"猫柠猫，这是我的好朋友火烈鸟，让他带我们走！"

他们乘着火烈鸟(12)越飞越高，火烈鸟美丽的粉红色羽毛好像天边的彩霞。他们飞过了亚马逊大王莲(13)，圆圆的莲叶比猫柠猫睡觉的窝还大。他们在树林中停歇，树上挂满一颗颗两头尖尖的大果实。

"这是可可树(14)，尖尖的果实是可可豆，可可豆做出的饮料苦里带着香，很好喝！"

"原来这就是可可豆，在我们那里，会用可可粉做巧克力！"猫柠猫说。

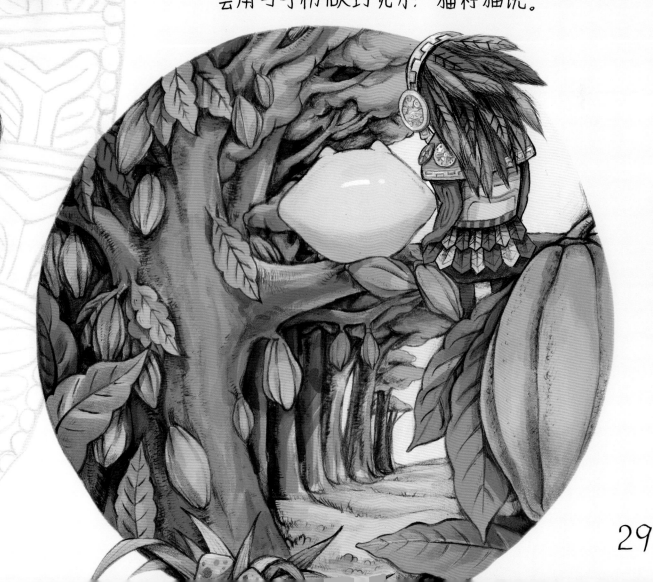

最后,他们飞到一个宏伟的建筑物的上空,这是一个上小下大的建筑,顶部有一个平台。

玛雅自豪地介绍说:"这是我们祭祀用的建筑——玛雅金字塔(15),上面雕刻着我们的文字,记录着我们的文明,这里是我们最神圣的地方。"

Finally, they arrive at a triangle-like building with pointy top but big bottom. There is a platform on the top of the building.

Maya proudly introduces "this is a Mayan Pyramid (15), the building we used to perform sacrifices. There are our calligraphy carved on the buildings. Our culture has also been well preserved in these buildings. This is the most sacred place around here."

猫柠猫仔细看着建筑物顶上的文字,好像自己的宝书《文字秘籍》上的符号呀!他明白了:原来这些符号是玛雅文!

他突然又想到,既然这些玛雅文能带自己来这儿,一定也能带自己回去!

"玛雅,把我放到平台上吧,这些文字也许可以把我带回家!"

Morningmore looks closely. The glyphs carved on the building look just like the symbols on the "Mystical Book." "Oh, I see. Those symbols are Maya script."

He wonders if those script brought him here, they definitely can bring him back home. "Maya, take me to the platform. Those scripts might take me back home."

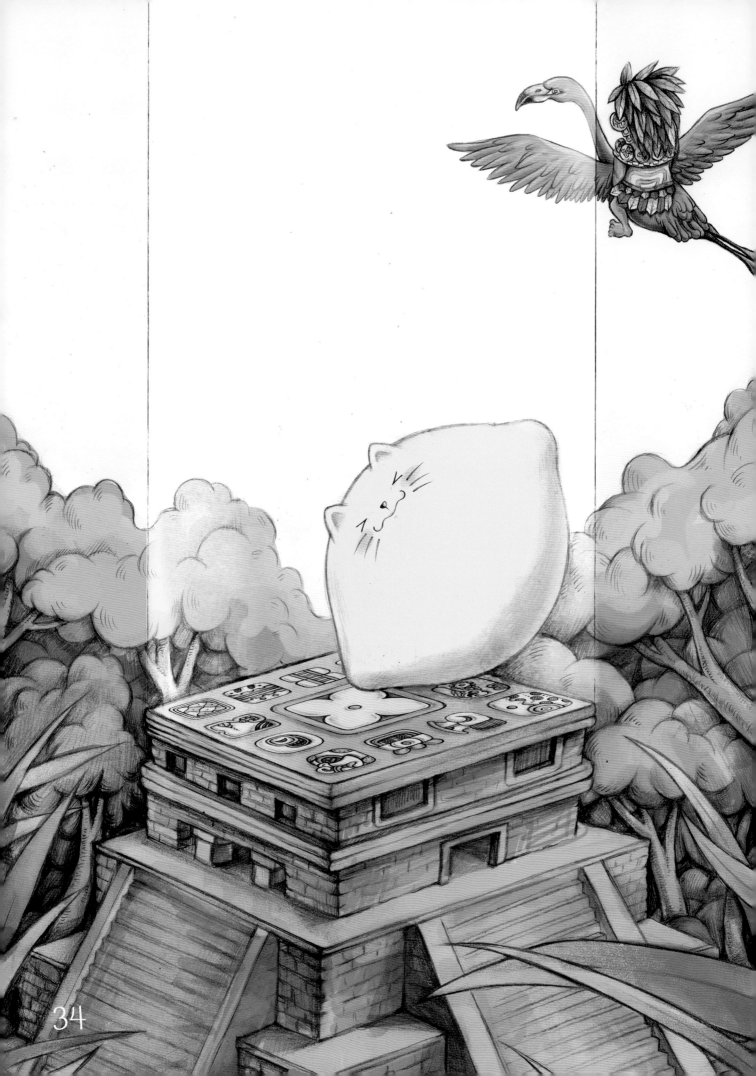

猫柠猫跳到了玛雅金字塔顶,他伸手去摸那个他来时摸过的符号,符号立刻放出金色的光芒。猫柠猫回头对着玛雅喊道:"谢谢你!玛雅,再见啦!"

下一刻,金色的光把猫柠猫包围,刺得他睁不开眼。

Morningmore jumps off onto the top of the Pyramid. He reaches out to the same symbol that brought him here. Suddenly, the symbol shines with golden glow. Morningmore turns back, "Oh no, I will miss you! Thank you so much for your help!"

Immediately, the golden glow surrounds Morningmore, making it impossible to open his eyes.

当猫柠猫再次睁开眼睛的时候,他发现自己又回到了绘本馆里。一切都和之前一模一样,一切好像一场梦。

猫柠猫把《文字秘籍》收好,藏回了书架。但他睡不着觉,一直在想:"为什么玛雅文会把我带到美洲丛林呢?"

(答案下页揭晓)

Morningmore realizes he is back to the shop when he opens his eyes again. Everything comes back to normal. It's just like a dream.

Morningmore puts away "Mystical Book" and hides it carefully. He can't fall asleep and wonders why the "Mystical Book" takes him to American Jungle.

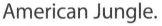

本期故事小问答:

◉ **提问：为什么猫柠猫碰到玛雅文会穿越到美洲丛林呢？**

Why the Mayan Script takes Morningmore to American Jungle?

回答：玛雅文来自2000多年前的一种文化——玛雅文化。该文化在丛林里孕育和成长，丛林就是玛雅文化的"家"。所以，当猫柠猫触摸玛雅文，就会来到这些文字的故乡——美洲丛林。

Mayan culture is a culture dated back to over 2000 years ago. It's a culture cultivated and developed in the rainforest. As a result, when Morningmore touch the Mayan symbol, it brings Morningmore to the birthplace of the culture, the American Jungle.

红带袖蝶
Postman butterfly

原来如此啊！

- **故事中的三种蝴蝶，你找到了吗？**
  答案在这里！更详细的介绍在后面哦！

光明女神闪蝶
Morpho Helena

88 多涡蛱蝶
Anna's eighty-eight

猫柠猫秘密图鉴
Morningmore's secret archive

01 大根乃拉草
Gunnera manicata

原产于南美洲巴西亚马逊河流域一带的大型观叶植物,是世界上目前发现的草本植物中叶片最大的植物之一。
Known as giant rhubarb, is a species of flowering plant in the Gunneraceae family from Brazil. It is a large, clump-forming herbaceous perennial.

02 美洲丛林
American Jungle

位于美洲的热带雨林,主要分布于亚马逊河流域,物种繁多,降水丰富,其中面积最大的是亚马逊热带雨林,有"地球之肺"之称。
Mainly located around the Amazon River. It is the most species-rich biome. The Amazon rainforest is the biggest among them and it is called lungs of earth.

03 行走棕榈树
Socratea exorrhiza

广泛分布于中美洲和南美洲。据说能自行移动,其复杂的根系可充当"腿",当季节变化时,可帮助其向有阳光的地方移动。
Known as the Walking Palm, is a palm native to rainforests in tropical Central and South America. There's a myth that they can walk with their roots following the sun as seasons change.

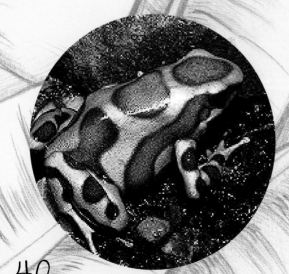

04 箭毒蛙
Poison dart frog

世界上外表最美丽的青蛙,同时也是毒性最强的物种之一。它们的体型很小,最小的仅有1.5厘米,个别种类也可达到6厘米。
The most beautiful frog in the world and one of the most poisonous species at the same time. Most species of poison dart frogs are small, sometimes less than 1.5 cm in adult length, although a few grow up to 6 cm in length.

05 蜂鸟
Hummingbird

世界上最小的鸟类，大小和蜜蜂差不多，由于它飞行采蜜时发出嗡嗡的响声，因而被人称为蜂鸟。飞翔时两翅急速拍动，频率可达每秒50次以上。
They are birds native to the Americas, among the smallest of birds. They are known as hummingbirds because of the humming sound created by their beating wings which flap at high frequencies, which vary from around 12 to 80.

06 88多涡蛱蝶
Diaethria Anna

属于凤蝶科的一个物种，分布于南美洲，约有40余种品种。因下层翅膀上的"8"字型图案而得名。曾被评选为"世界上最美丽的八种蝴蝶"之一。
Commonly called Anna's eight-eight, in reference to the characteristic patterns on the hindwing undersides of many. In Diaethria, the pattern consists of black dots surrounded by concentric white and black lines, and typically looks like the numbers "88" or "89".

07 光明女神闪蝶
Morpho Helena

鳞翅目，蛱蝶科，闪蝶属的一种蝴蝶。在热带雨林出没，也适应南美气候。雄蝶有领域性，翅膀反射出的金属光泽是在向其它雄蝶表示其领域范围。
It is found in the rainforests of northern South America. It is known for its metallic blue and shiny wings. The wingspan is 75 to 100 millimeters.

08 红带袖蝶
Heliconius Melpomene

又名红色邮差蝴蝶，属于凤蝶科的一个物种，主要分布在中美洲至巴西南部地带。为了降低遭袭击的可能性，通常会四五只聚集在一起过夜。
Known as postman butterfly. The postman butterfly is predominately black and either red or yellow stripes across the forewings. It is found from Mexico to South America. Individuals of postman butterfly for large communal roosts which they return to each night after foraging, which provides protection from predators.

09 金刚鹦鹉
Macaw

产于美洲热带地区，是色彩最漂亮艳丽的鹦鹉，也是体型最大的鹦鹉。
Native to Central America and North America, South America. Proportionately larger beaks, long tails and relatively bare, light-colored, medial areas distinguish macaws from other parrots.

10 凯门鳄
Caiman

亚马逊河里凶猛的动物。分布于从墨西哥南方到巴西的热带地区，外表与美洲鳄鱼相似。
Fierce animal living in the Amazon River. Due to the large size and ferocious nature of the caimans, they have few predators within their environments.

11 火烈鸟
Flamingo

火烈鸟亦称红鹳。身体羽毛白色、玫瑰色相间，翅膀羽毛黑色、深红色相衬，非常艳丽。
A type of wading bird in the family Phoenicopteridae. With their pink and crimson plumage, black wings and beaks, flamingos look flamboyant.

12 亚马逊大王莲
Victoria Amazonica

大王莲又名维多利亚睡莲，是世界上最大的水生有花植物之一，原产于南美洲的亚马逊河。是圭亚那的国花。
A species of flowering plant, the largest of the Nymphaeaceae family of water lilies. It is the National flower of Guyana.

13 可可树
Theobroma cacao

原产于热带美洲，其果实经过发酵及烘焙后可制成可可粉，是玛雅人最为重要的经济林木。三千年前，美洲的玛雅人就开始培植可可树。
Also called cacao tree. It is a small evergreen tree in the family Malvaceae, native to the deep tropical regions of the Americas. Its seeds cocoa beans, are used to make cocoa mass, cocoa powder and chocolate.

14 玛雅金字塔
Maya pyramid

玛雅文明的象征，玛雅金字塔和埃及金字塔不同，外型上，玛雅金字塔是平顶的，有层叠的台阶，塔顶的台上建有庙宇。功能上，玛雅金字塔主要用来举行宗教仪式。
Symbolization of Mayan culture. Maya pyramids are different from Egyptian pyramids. Architecturally, Maya pyramids are flat-roofed with steps on the side and temples sat atop. Functionally, Maya pyramids are mainly used for performing sacrifices.

注：以上内容来自网络